BRITISH PIERS

BRITISH PIERS

Photographs by Richard Fischer

Introduction by John Walton

Thames and Hudson

First published in Great Britain in 1987 by
Thames and Hudson Ltd., London
© Edition Braus, Heidelberg
Photographs © Richard Fischer, 1987
Printed and bound in West Germany, Brausdruck GmbH, Heidelberg

CONTENTS

PHOTOGRAPHERS NOTE

I should like to thank all those who have helped me un-officially and informally. It is always refreshing to find how relaxed and cheerful the British are when con-fronted by a foreigner with a camera prying into their lives. I dedicate this book to all those who have been in-volved in creating it, and especially to my wife Renate, who so patiently put up with the difficulties and shared in the problems that we encountered on our journeys.

I used various cameras during my four trips to England:
2 MINOLTA XG-M bodies with 20 mm, 35 mm, 100 mm and 200 mm Rokkor lenses
and
TOYO FIELD 810 M 4 x 5" and 13 x 18 cm. Large format with Nikkor 75 mm, Rodenstock 250 mm and Nikkor 450 mm lenses.
All the photographs were taken on Kodachrome and Ektachrome film, and occasionally graduated filters were used for more dramatic effects.

Richard Fischer

Photographic Acknowledgments

p. 9 Avon County Council, Woodspring Central Library.
p. 11 Kent County Library, Local Studies Collection, Margate Library.
p. 15 East Sussex County Library.
p. 17 Southend Central Library.
p. 21 North Yorkshire County Library.
p. 23 Blackpool District Libraries.
p. 25 East Sussex County Library.
p. 27 East Sussex County Library.
p. 29 Cleveland County Libraries.

INTRODUCTION

The seaside holiday was one of the happiest inventions of the Industrial Revolution years in England, and in Victorian times the seaside pleasure pier became a popular and evocative aspect of it. Unlike many of the more workaday or mundane products and innovations of British industry, however, the seaside holiday, as such, took a long time to spread beyond the confines of these islands; and its growth was stunted even in Scotland and Ireland. The most successful seaside resorts were prominent among the fastest-growing English towns during the second half, and especially the last quarter, of the nineteenth century; but only on parts of the eastern seaboard of the United States did development take place on anything like a similar scale, with Atlantic City conspicuous in the vanguard. Otherwise, seaside resorts grew up as part of an expatriate British culture, or catered for the needs of those British holidaymakers who crossed the Channel in search of cheaper and more exotic pleasures than prevailed at home. Other cultures took less readily to the beaches and the sea on pleasure bent, and the European seaside had a lot of catching up to do in the inter-war years and afterwards while, characteristically, the English resorts languished and failed to adapt.

The seaside pleasure pier proved much more insular and specific to a particular culture. On a notional scale of the general popularity and exportability of British leisure innovations, of which there have been a remarkable number, the pier falls closer to fish and chips, which have never caught on outside the British Isles, than to football, with its almost world-wide success. It would be interesting to speculate on why this has been so: on why the pleasure pier has been such a narrowly British, and indeed English (and Welsh) phenomenon. But, more importantly, we need to describe and explain the emergence and spread of the pleasure pier in its native land, and to account for the declining importance of the pier in the seaside resorts of the twentieth century.

The pleasure pier is often popularly supposed to be an essential adjunct to the Victorian and Edwardian seaside holiday. This is an exaggerated view. Not all resorts acquired a pier, and the pier is less prominent than the bathing-machine, the donkey or the rapacious seaside landlady in what might be called the folklore of the seaside. In the *Punch Library of Humour* compendium of Victorian and Edwardian seaside jokes, for example, the pier plays a very occasional and secondary role. Nevertheless, from the pioneering Brighton Chain Pier of 1823 to the opulent Edwardian speculations in the popular resorts, the coastline of England and Wales became irregularly punctuated by pleasure piers of varying length and ostentation, with peak periods for development coming in the 1860s and the 1890s. A national resort guide of 1895 provides evidence of at least seventy pleasure piers in England and Wales; and although some were soon to be lost to rough seas and collisions with vessels, more were also added, both at small and aspiring resorts like Minehead and Fleetwood and as additional attractions at established centres such as Weston-super-Mare and Brighton. The geographical distribution is revealing. The most obvious concentration was in the north-west, and especially in Lancashire, where Blackpool, the pioneer working-class resort, already had three pleasure piers, Morecambe was about to acquire a second, and Southport, Lytham and St. Anne's had one each. Just across the Mersey, New Brighton also had a pier, and the North Wales resorts had three more. This high level of provision reflected the proximity of this coastline to the great industrial centres of Lancashire, the West Riding of Yorkshire and the West Midlands, which provided both customers and capital for pier investment, and stimulated competition to provide artificial attractions between resorts whose accessibility and natural endowments were not strikingly distinctive or at variance. The area also supported an intensive network of pleasure steamer services, which provided an added incentive to investment.

The Isle of Wight also positively bristled with piers. Seven were listed in 1895, at obscure places like Sea View and Totland Bay as well as more obvious holiday centres like Ryde, Shanklin, Sandown and Ventnor. Here the transport function was more important than pleasure, especially in the early days of the island's railway system. Steamers were the lifeblood of the small resorts, and the proliferation of piers was probably a very direct response to the competition for a limited number of visitors.

Yorkshire's long coastline also sprouted seven pleasure piers in 1895, but their importance and impact were much less marked than the Lancashire concentration. The rival piers at the twin resorts of Redcar and Coatham, on Teesside, and Saltburn pier not far away, were aimed at attracting custom to remote resorts which were being speculatively promoted; and at the other end of the county, similar things might be said of Hornsea and Withernsea, which had little else to offer in the eyes of unkind contemporaries. Only Scarborough among the major resorts had acquired a pleasure pier as such: Whitby, Filey and Bridlington managed without.

Practically all of the remaining pleasure piers were spread more evenly along the east and south coasts, from Cleethorpes in north Lincolnshire to Plymouth in Devon. Some of the larger resorts boasted more than one: Great Yarmouth, Brighton and Southsea had two each, and Bournemouth had three if we include its satellite resorts of Boscombe and Southbourne. There were concentrations in Essex and Sussex, where the pull of London, still the biggest source of holiday demand, was strongest; and the Essex resorts, especially, needed piers to give them access to the Thames estuary steamers which brought a large proportion of their visitors. New resorts without harbours, such as Clacton and Walton-on-the-Naze, were particularly dependent on piers as transport arteries.

On the other hand, there were extensive tracts of coastline which were almost innocent of piers. Cornwall was being discovered by discerning holidaymakers, but its amenities did not extend to pleasure piers: most of its

visitors were self-consciously pursuers of the rustic, romantic and bucolic, and there was little to tempt pier builders on a generally inhospitable coastline. The mud flats of the Bristol Channel and the population centres of Bristol, South Wales and the Midlands encouraged half a dozen piers to cater for promenaders and steamer passengers; but from Swansea Bay to Beaumaris the rest of Wales knew only a single pier, at Aberystwyth. As in Cornwall, the resorts were too small and remote, and their visitors too few and too uninterested in artificial entertainments, for piers to be worth building. The same applied to the far north-west, where the only pier north of Morecambe was the speculative venture at Silloth, where the efforts of Carlisle businessmen to found a commercial port and resort bore limited fruit.

Piers, then, were not synonymous with the Victorian seaside holiday: they were associated above all with a distinctive kind of Victorian seaside holiday, the bourgeois, family-centred, sedate, conformist, culturally undemanding fortnight or month at the coast which became, and remained, the mainstream holiday market upon which most resorts depended. Where the working classes obtained a foothold in the later nineteenth century, piers, like other forms of entertainment, might be driven down-market and become associated with more raffish and proletarian pleasures; but this remained exceptional. The geographical distribution of piers in their late-Victorian heyday is partly a matter of topography: open shores and estuaries provided the most plausible and practical pier-building sites, in contrast with the rocky coves and difficult terrain of the most scenic bits of coastline. But the long beaches were also more attractive and accessible to the kind of visitor who liked bathing-machines, sandcastles and piers rather than 'roughing it' in pursuit of quiet and romantic scenery, and there were social as well as topographical reasons for the pattern of pier-building around the coastlines of Victorian England.

The figures I have been using are approximate, because there is no clear dividing-line between the pleasure pier and other kinds of pier, especially as most

Birnbeck Pier, Weston-super-Mare,
linking the shore with an island: an
engraving dating from soon after its
completion in 1867.

pleasure piers also derived revenue from passenger transport and even the handling of cargo. To qualify as a 'pleasure pier' for present purposes, a pier must have been purpose-built, usually of cast iron and wood, to provide a promenade and other amenities (if only fishing and the occasional band) as well as acting as a landing stage. The many 'piers' which were built of stone to act as commercial quays, harbour walls or sea defences are not included in this definition, although many were used as promenades and sometimes a toll was charged. At Bridlington, for example, the harbour was enclosed by stone piers to north and south, and the 'north pier, lighted by lamps, forms a very pleasant and largely patronised promenade'. At Watchet, on the Bristol Channel, a breakwater was used as a promenade, and at Barmouth the long railway bridge across the Mawddach estuary performed the same functions as a promenade pier. So the definition of a 'pleasure pier' is not as clearcut as one might think, and this complicates attempts at counting and assessing patterns and trends.

Most pleasure piers looked distinctive, however. They pointed determined fingers into the sea, sometimes with elaborate pavilions and other buildings at the shore end, sometimes with wider platforms and ornamental buildings at the sea end, and often with kiosks and shelters lining the main deck. They were designed by engineers rather than architects, and much of the elaborate decoration of the ironwork on their superstructure was chosen from ironfounders' catalogues. But they were remarkable structures, with their intricate spidery superstructures and sturdy supports, and the Moorish or eastern idioms in which many of their associated buildings were embellished. Some diverged significantly from the pattern outlined here: Weston-super-Mare's Birnbeck Pier linked the shore with a conveniently-placed island, on which amusements were eventually developed, while Southsea's Clarence Pier 'is large and spacious, does not run out into the sea, but lies along the shore, upon it being a splendid pavilion in which military bands play daily.' Most elegant of all were the piers built on the suspen-

sion bridge principle, the Brighton Chain Pier and the similar structure at Sea View, Isle of Wight. Not that elegance was inevitable, or universally perceived: an acid *Punch* contributor must have expected to find sympathy among his readers when attacking, in writing about an unnamed resort, 'that ridiculous apology for a pier, with its long, lanky, bandy legs, on which I have been dragged every evening to hear the band play'. Some piers had tramways, some had lateral extensions to accommodate steamers, and many kept acquiring additional buildings and amenities in the late nineteenth and early twentieth centuries. For all their variations and occasional idiosyncracies, their differing lengths and outlines, they formed a characteristically Victorian building type whose construction, purpose and use says a great deal about the nature of the culture which produced and maintained them.

How did the English pleasure pier originate, and what was the chronology of its development? As a promenade, it was a seaside variant of an established amenity of the spas and centres of county society in the eighteenth century. Here, the development of tree-lined gravel walks, which were kept socially select by tolls and regulations, had been a response to the need of an increasingly competitive polite society to see and be seen in a formal setting. Resorts like Bath and county towns like Preston brought together the aristocracy, gentry and aspiring middle ranks, professionals, merchants and the like, in an increasingly complex social system. The older hierarchical view of society, in which one's place in the world was inherited and almost unchallengeable, was giving way to a system in which status was acquired by the display of opulence and social graces, as well as being ascribed on the basis of birth and inheritance. The rise of the 'leisure towns' of the eighteenth century was part of this process, and they provided arenas for competitive display and the formal, conventionalised flaunting of wealth and urbanity. The development of the seaside resort from the mid-eighteenth century followed this established pattern, with the introduction of formal promenades and as-

The first Margate Pier, built in the
1820s by Rennie, was really a stone
jetty. This was often the origin of the
iron piers built later in the century,
in the case of Margate about 1855.

semblies in which polite, regulated social interaction, which included controlled flirtation and the operation of a marriage market, could take place. When technological innovations enabled select promenades to be extended over the sea as well as to skirt its edges, the promenade pier became an almost irresistible option. Not only did it provide a readily segregated and controllable promenading space: it also provided enhanced access to sea air, the curative properties of which were held to be second only to those of sea water; and it allowed promenaders to look back at the shore, giving them picturesque inland views which gave the illusion of being on board ship without the discomfort. On top of this, of course, pier promoters could enhance their toll revenue by collecting embarkation and landing dues from boat passengers; and in the early days, especially, this was the most lucrative aspect of pier finances.

The first pleasure piers were opened, fittingly, at Margate and Brighton, the first English seaside resorts to combine fashionable allure with popularity and grow to a substantial size. The rebuilding of Margate's landing pier after storm damage in 1808 included a gallery where a band played, and there was a promenade, to which a penny admission fee was charged. This innovation was greeted by riots when it was introduced in 1812, and one of the toll gatherers was nearly thrown into the sea. It would be interesting to know more about these events; but they were no doubt connected with the increasingly socially mixed nature of Margate's visiting public, which by this time included a large number of decidedly unpretentious Cockney tradespeople and shopkeepers. Margate's pier rebuilding was more a matter of the addition of promenading and entertainment facilities to an existing structure, than the construction of a purpose-built pleasure pier, however; and Brighton's Chain Pier was probably the first of the new breed. It was planned as a convenient landing stage for continental passenger traffic, as well as being a fashionable promenade; but the company's optimistic prospectus in 1821 anticipated a considerable return from the latter source, and a twopenny charge was levied on

strollers. The pier opened in 1823 after severe construction problems had been overcome, as the work was 'constantly interrupted by storms'; and the eventual cost was £30,000.

The Chain Pier's finances were somewhat chequered, but it became a valued social asset to the town, as well as a port of call for steamers and an attraction for sightseers. It was a vantage point for invalids to inhale the sea air, and it became a favourite source of gentle exercise for King William IV. It soon acquired amusements: a regimental band, telescopes, a camera obscura, a saloon and reading room, souvenir shops and kiosks, a silhouettist, a weighing machine, regular firework displays and 'meteorological prognostications'. In this it was a long way ahead of the piers which soon began to appear in rival resorts; but this was fitting, for Brighton was very much the fashionable pacesetter among English seaside resorts from the late eighteenth century to the 1840s, and it had a uniquely numerous and opulent visiting public.

The piers of the second quarter of the nineteenth century had, indeed, few pretensions as promenades. Their overriding purpose was to ease the way for the seaborne visitors whose security and comfort were of such competitive importance to resorts before the development of the railway system. Most, if not all, were wooden constructions, and they were concentrated into south-eastern England, where rising resorts were vying for the London steamer traffic along the Thames estuary. Gravesend and Herne Bay acquired piers of this kind, and the most spectacular case of pier promotion on this model was at Southend. Here, a syndicate obtained power to build a very long pier of a mile and a quarter in 1830, although the full scheme was not completed until 1846, and there was a frustrating hiatus while steamers were only occasionally able to use the half-mile section which had been opened. The pier was a financial disaster, and eventually had to be bailed out and rebuilt by the local authority, as we shall see; but its limited early impact on the resort's prosperity is suggested by *Punch's* satire in 1848, when Southend was said to be 'a mere shrimp of a sea-town', almost un-inhabited, silent as the grave, and 'well worth seeing – with a microscope'. This is a reminder that the piers of this period were far from being golden keys to inevitable prosperity for the towns at which they were built, or in-deed for their promoters; and this remained the case during the great acceleration in pleasure pier specula-tion which arose during the 1850s and especially the 1860s.

What really opened out the mid-Victorian pier-building boom was the increased availability of limited liability for shareholders in speculative companies after 1856 and 1862, coupled with the improved construc-tion methods for iron structures and piledriving which became available in response to the needs of the rail-ways. These opportunities were rendered irresistible by a concurrent rapid expansion in the middle-class sea-side holiday market, at a time when the railways were opening out new resorts and intensifying the competi-tion between existing ones. This tempting combination of reduced risks, lower costs and visibly growing de-mand lured company promoters and investors into sub-stantial speculation in pleasure piers in a wide range of resorts on most of the coastlines of England and Wales, while it was obviously dangerous for any aspiring resort to sit idly by and watch its competitors acquire attrac-tions which might give them the edge in the competition for growth and prosperity.

The pier boom of the 1860s can be made to seem of little account compared with what followed. Eric Hobsbawm has calculated the amount of 'projected in-vestment in pleasure piers' as averaging £36,000 per year in 1863–5 and £27,500 per year in 1866–70, com-pared with over £70,000 in the early 1880s and well over £80,000 through the 1890s. But this was projected rather than actual investment, and many schemes never came to fruition, especially in the later years when out-side speculators were moving in on what appeared to be a lucrative market. An unstated proportion of the in-vestment, moreover, was on ordinary harbour works. The important thing about the 1860s is that most of the proposed pleasure piers actually were built, and the spate of new constructions marked the real beginning of the promenade pier as seaside institution. The relative smallness of the sums invested also masks the fact that a large number of cheap, basic promenade piers could be obtained for quite a limited outlay, to be embellished at greater expense later on; and this is certainly what happened. Taken as a whole, investors produced more piers for their money in the 1860s than in the 1880s or 1890s. At least twenty-four were constructed during the decade, and several more were well advanced in 1870. The south coast resorts, from Folkestone to Bourne-mouth, were especially prolific, as was the arc of north-western coastline from Rhyl to Morecambe.

Almost all of these piers were promoted and funded by local residents in the resorts themselves, or by businessmen and holiday visitors from the surrounding area who could be expected to identify with the resorts and to know them well. Shopkeepers, tradesmen, sea-side landladies and professional men were well to the fore, and in some resorts, especially Southport, working-class people risked their savings in pier company shares. They included fishermen, domestic servants, a waiter and a barmaid. Such people could be expected to know what their visitors wanted, and to profit indirectly from a general stimulus to the holiday season, as well as directly from possible dividends on their shares. They were keen to improve the competitive position of their own resorts, and also to pre-empt any outside speculators who might seek to funnel the profits from a successful pier into their own hands.

Events at Scarborough in 1864–5 illustrate this theme. In 1864 a syndicate of Manchester businessmen sought powers to construct a promenade pier from a point close to the harbour. At once, a storm of local op-position was unleashed, led by a cross-section of the town's leading citizens. They argued that the proposed pier would interfere with navigation into and out of the harbour, and would actually be dangerous to shipping. There was also a threat to amenity and morality, as promenaders on the pier would be able to observe bathers on the south sands from a strategic height, at a time when male bathers at Scarborough were still not being required to wear bathing drawers. The promoters responded by asserting that problems of access to the harbour would only arise under certain unusual combinations of wind and tide, and offered to meet the modesty problem by erecting a movable screen, six feet high, which could be adjusted according to the state of the tide in order to frustrate prurient promenaders. They also alleged that their opponents, who described themselves as 'the merchants, bankers, shipowners and other inhabitants of Scarborough', had vested interests to protect, as they were shareholders in a rival concern, the Cliff Bridge Company, which provided pleasure grounds and promenades on the shore. The ultimate arbiters in the dispute were the officials of the Board of Trade, the government department responsible for the protection of shipping and of the Crown's interest in the foreshore; and when they decided that the pier would be a threat to navigation, and could not be shortened or adjusted in such a way as to remove the threat, the promoters were unable to proceed. But the Manchester businessmen were probably correct in identifying the vested interests of their opponents; and their most cynical suspicions must have been confirmed when in 1865 a syndicate of Scarborough bankers and merchants successfully promoted a pier on the north sands, at a safe distance from the harbour, which was duly constructed, although it was not financially successful in the long run.

This episode should also remind us that pier promotion was not a straightforward matter. Consent had to be obtained from Parliament and from the relevant foreshore authority, which was not always the Board of Trade: it might be the Office of Woods and Forests, or the Duchy of Lancaster, and the Lord of the Manor might also have foreshore rights. The government departments often seem to have fought against each other rather than collaborating to make administration easier, and there were dangerous administrative pitfalls. The Cleethorpes Promenade Pier Company of 1866 jumped through all the bureaucratic and legal hoops, only to find that the Board of Trade's approval was irrelevant, as the pierhead was outside the Board's jurisdiction, and was the responsibility of the Conservancy Commissioners of the River Humber. This mistake held matters up and must have involved the Company in additional legal costs. Building expenses also had a habit of escalating, despite the development during the 1860s of prefabricated piers which could be assembled on the site, as was done at Clevedon, in the Bristol Channel, and at Morecambe, where the components had originally been destined for Valparaiso. But the cost of Weston-super-Mare's Birnbeck Pier, to give an extreme but telling example, was increased severalfold to a crippling £70,000 when tidal problems were encountered; and pier speculation was undoubtedly a much riskier business than many of its devotees can have appreciated. The vulnerability of piers to storm, collision and fire was to become apparent later in the century, as we shall see; and at Bournemouth the town's first pier, built for the local authority in 1859–61, was already suffering before the decade was out. It was one of the last of the all-wooden piers, and it proved disastrously attractive to the tastebuds of the teredo or shipworm. Its ravages were already apparent by 1867, when the pierhead was destroyed by a gale; and the pier's fate was sealed by the loss of a further length of woodwork in 1876.

The piers of the 1860s were narrow in width, with few buildings to clutter their promenade decks, and little scope for any activity other than taking the air and enjoying the view. Southport's pier was only sixteen feet wide when first built, even though it carried a 'tramroad with a single carriage, propelled by manual labour'. It was widened to 24 feet in 1864, when a fixed engine, an endless rope and improved carriages were ordered for the tramroad. The North Pier at Blackpool was 28 feet wide, although the pierhead measured 135 feet by 55 feet. The length of these piers was often much more impressive. Southport's was 1200 yards long when first built: it is not surprising that it depended on the new technologies devised for building extensive railway via-

The Chain Pier, Brighton, most
famous of the early piers. Opened in
1823, it was destroyed by a fierce
storm in December 1896.

ducts across the deep indentations of Morecambe Bay. In 1863 it was extended by a further 265 yards to reach low water mark, and the extension was assimilated to the main body of the pier in 1867. Southport thus became second only to Southend in the league table of longest piers; and its contemporaries were less ambitious, perhaps partly because the sea was not so far away in other resorts. Most were between 300 and 500 yards in length, and some were considerably shorter. Even so, these were conspicuous and distinctive structures; and in terms of overall proportions, the piers of the 1860s set the pattern for future developments. Subsequent piers were often wider, at least in part; but the main changes in succeeding decades, which applied both to new and to existing piers, involved the addition of pavilions, lounges and other more ambitious buildings to the existing basic structure.

The promenade piers of the 1860s might be limited in the scope of the entertainment they provided, but they aroused high expectations and even heady excitement when they opened for business. They were seen locally as harbingers of unexampled prosperity. There were exceptions: the West Pier at Brighton was controversial. The inhabitants of Regency Square, a high-class residential development which overlooked the pier entrance, complained that the toll-houses were an eyesore which obstructed their sea view. This kind of sectional and local objection was not uncommon later in the century; but the typical response of seaside communities in general is well expressed by the festivities which greeted the opening of the first pier at Blackpool in 1863, when the resort in question was still small, provincial and unremarkable. This description comes from the Pier Company's golden jubilee souvenir, so the lily may well have been gilded; but it conveys the spirit of the occasion convincingly enough:

'Tradespeople closed their shops; householders made their places of abode gay with flags and bunting; and all classes of the community entered fully into the festive spirit. The only piece of artillery of which Blackpool could then boast – a small twelvepounder cannon . . . boomed forth in salutation of the advent of the new era of enterprise in Blackpool's history. And . . . there was a grand procession, to the martial music of several brass bands, of the Volunteer Artillery and Rifle Corps, of Freemasons and Friendly Societies, Fishermen and Bathing-machine Attendants, Trades and Labour Societies, Day and Sunday School Children, Representatives of the Local Governing Authority, and Directors and Shareholders of the Pier Company, all of whom were animated with the one object – "to mark the day as noteworthy".'

As it turned out, Blackpool's first pier was indeed the initial link in a chain of successful innovations which underpinned the resort's extraordinary expansion in the late nineteenth and early twentieth centuries. It was also a financial success, with dividends hovering steadily around 12 per cent for most of its first fifty years. This was excellent going by any standards, even those of Blackpool in its boom years; and elsewhere the rewards for shareholders were less lucrative. In the mid-1870s successful piers were paying between six per cent (Eastbourne) and ten per cent (Hastings) in dividends, with Morecambe, Southport and the West Pier at Brighton coming between these respectable figures. But for every success story there was a failure. Weston-super-Mare's Birnbeck Pier paid no dividend for its first seventeen years; Saltburn's pier was partially demolished by a steamer in 1875, while the town's economy languished and empty houses proliferated; and even Scarborough was unable to sustain a profitable promenade pier, as the company formed in 1865 had to be wound up in 1889 after years of financial difficulty. The most successful of the first wave of promenade piers were those which were able to adapt to the changes in the holiday market which were already becoming apparent in the 1870s, and diversify their attractions to cater for working-class trippers as well as middle-class families.

Most of the piers of the 1860s charged considerable sums, in the context of the time, for admission to their limited attractions. This ensured a measure of social

Southend Pier.

The first Southend Pier, begun in
1830 and extended in 1845. By
1872, the date of this engraving,
horsedrawn trams carried passengers
to the landing-stage at the end.

selectness, filtering out undesirables; and this exclusiveness was itself an important attraction, for it was not obtainable on the beach or in the street, and only short stretches of sea-front promenade, in a few resorts like Bridlington, were policed in this way. The West Pier at Brighton charged as much as sixpence, and became 'the promenade for visitors with social aspirations'. In 1875 more than 600,000 admissions were recorded; but in no other British resort, even at the height of mid-Victorian snobbery and fear of contamination by the lower orders, were so many prepared to pay so much for so little. The Cleethorpes Pier Company, conscious of the persisting importance of the Lincolnshire gentry to the resort's fortunes in the mid 1860s, aspired to a fourpenny toll for promenaders and sixpence for bath- and sedan-chairs; but these socially ambitious intentions were thwarted by events. Blackpool's North Pier was closer to the mainstream in charging twopence for admission, but it was only able to preserve a modicum of selectness because working-class demand was channelled southwards to the town's second pier, the South Jetty (now the Central Pier), which opened in 1868. Under the pressures of changing demand patterns, coupled with the threat of competition from new entertainment centres on the shore, the pier in the popular resorts had to diversify their attractions, and the promoters and directors of new piers which were constructed in the later nineteenth century often had to take account of the new pressures and opportunities.

Cleethorpes provides a poignant example of a rapid change in expectations and provision. Aspirations to gentility were swiftly put aside, and by 1876 the pier's promenading function was overshadowed by the provision of a dance band for the armies of excursionists from the Yorkshire industrial towns. Residents complained of 'unseemly' conduct, with 'young men dancing together, smoking small black pipes', and alleged that the resulting influx of the 'great unwashed' prevented 'respectable' people from enjoying the promenade, robbed Cleethorpes of what had been its greatest charm, and 'stamps the pier as a resort for a class of people for whose "amusement" it was surely never intended'. The dancing was subsequently transferred to a separate concert hall, which also provided music-hall entertainment; but the pier's down-market posture was inescapable as its resort inexorably became a working-class playground and fairgrounds dominated the beach.

Blackpool, which was beginning its evolution into the world's first fully-fledged working-class seaside resort, was encountering similar problems at a slightly earlier stage. Despite its twopenny admission charge, the North Pier was swiftly inundated with trippers at Whitsuntide and in July and August: it was claimed that 'respectable visitors would not go upon the pier during the time that the excursionists were there'. The South Jetty was partly a response to this problem, promoted by a dissident group of North Pier shareholders. It was strategically placed opposite the popular Wellington Hotel, whose proprietor cunningly donated a plot of land; it charged only a penny at the turnstile; and from 1870 onwards its cheap steamers, along with open-air dancing to the music of cheap German bands, ensured its popularity among the swelling throng of working-class visitors. The two piers divided the market between them, and during 1875–90 annual admissions to the South Jetty more than doubled from 481,000 to 988,000. Most of the newer pier's patrons were unpretentious members of the Lancashire working class: as a dialect writer put it, on the South Jetty 'th' bacca reech smells stronger, an' th' women are leauder abeaut th'meauth. We know what class their return tickets are . . .'

One or two other resorts also saw piers extending their range of attractions during the 1870s to embrace excursionists. Weston-super-Mare's Birnbeck Pier had 'swings, gymnastics and other games' by 1876, and in 1880 there were complaints that the band and other delights on Margate's recently completed Jetty Extension were inducing visitors to 'spend all the time on the Jetty Extension to the injury of the town' as a whole. Such innovations involved little in the way of additional investment, and the Weston pier seems to have lacked the capital to complete its refreshment pavilion, a much more demanding enterprise. From the 1870s onwards, however, a new phase in pier embellishment was beginning in earnest, as saloons and pavilions were added to existing piers and incorporated into the designs for new ones: and in this respect it was the piers which catered for the established middle- and upper-class visiting publics that led the way.

One of the pioneering concerns in this respect was Blackpool's North Pier, which in 1874 carried out a major scheme which increased its area by more than an acre, widening the deck and providing new north and south wings at the end. On one of these was placed an open-air bandstand, a suite of shops and a refreshment room; and on the other was the Pier's 'greatest glory – the Indian Pavilion . . . and being designed from studies of Hindu Temples and Palaces – principally the Temple of Binderabund – it is a typical example of Indian architecture'. Typical or not, it perpetuated a tradition of seaside orientalism in architecture which can be traced back to the Prince Regent's famous Pavilion at Brighton; but for some reason the inscription at the back of the stage, 'The Hearing falls in love before the Vision', was in Arabic, and this reinforces the impression that pier architecture really borrowed indiscriminately from a variety of exotic cultures and imaginations, and that pretensions to scholarship were uneasily worn. Not that this matters – as Sir Nikolaus Pevsner remarked of Eastbourne, it is above all 'jollity' that 'becomes a pier'. The point of the North Pier inscription, however, was to express an intention to provide the sort of music which would appeal to cultivated Victorian tastes in suitably socially select surroundings; and this was carried out by the introduction of a 35-piece orchestra, with well-known conductors and a one-shilling admission charge which was well-calculated to deter undesirables. The North Pier's heavy investment paid off admirably, and its example was widely followed in more straight-forwardly up-market resort settings.

Over the next ten years or so, many other piers acquired similar additions and pretensions. Llandudno's ambitious reconstruction provided space for a 40-piece orchestra and an audience of 2000 at the pierhead, and its new concert pavilion could seat 4000. Southend Pier was rescued from accelerating decay by the local authority, who paid £12,000 for it in 1875 and practically rebuilt it for about ten times that sum. The Pavilion cost £7000, with a 1200-seater hall which was leased to an entertainment contractor, and extensive refreshment rooms. There was also a famous pier railway. Lowestoft's pier extension was rebuilt after a fire in 1885 with a combined assembly room, reading room and refreshment rooms, surrounded by 'double promenade verandahs'. Examples could be multiplied, as the pier pavilion and its regular concerts became expected features of the holiday season at the larger resorts, especially those with aspirations to gentility or at least respectability. Even little Ventnor, on the Isle of Wight, whose pier was only 520 feet long by 22 feet wide, could boast a bandstand on the platform at the end, where the local Volunteers' band played three times a week. It was also 'amply provided with ornamental wood and glass screens, fully protecting from sun and wind', a reminder that one of the main preoccupations of mid-Victorian holidaymakers, and indeed Edwardian ones, was to avoid the slightest suspicion of sunburn. The cult of sun-bathing was an artefect of the inter-war years: hence the magnificent displays of parasols which feature on sunlit photographs of pier promenaders before the First World War.

New promenade piers were also being built in these years, without embellishment; but the second great boom in pier-building and pavilion construction came between the late 1880s and the mid 1890s. Existing promenade piers were adorned with pavilions at Eastbourne, Brighton, Worthing, Ryde, Sandown, Aberystwyth, Southport, Clacton, and Morecambe, while new piers including such opulent buildings were constructed at Folkestone, Hastings, Southampton, Boscombe, Torquay, Blackpool, Brighton and Morecambe again. This is not a complete list, but it gives some indication of the major surge of investment that took place in these years, transforming the hitherto sedate image of the seaside pier, and multiplying new attractions.

Why this nationwide interest in pier promotion and improvement? In part it was a response to perceived opportunity: the rapid growth in popularity of the seaside holiday was obvious to all in this period of falling commodity prices and buoyant real incomes among the lower middle and upper working classes. The chequered performance of British manufacturing industry at this time was an added incentive to speculation in entertainment, which looked glamorous as well as promising profit. This frame of mind encouraged the flotation of a wide range of other kinds of seaside entertainment company: Summer and Winter Gardens, Eiffel and Revolving Towers, music-halls, fairgrounds and assorted pleasure-domes. Many of these schemes left unwary investors with badly burned fingers, and some were fraudulent from the beginning; but some came to fruition, though few did well financially, and cumulatively they posed the threat of dangerous competition for the piers which had so often been the pioneers of capital-intensive seaside entertainment. Most of the expanded facilities at existing piers should be seen in the light of this threat of intensified competition, which applied to rivalries between resorts as well as within them, as well as being a response to widening opportunities. But the new piers of these years, some of which were the second (in Blackpool's case the third) in their resorts, offer a reminder that the pier in itself was far from being played out as an essential seaside attraction. They went far beyond the original provision of steamer access, promenading and the inevitable (but often occasional) band; but they added additional options on to this basic mid-Victorian menu, rather than displacing the older items or rendering them obsolete.

Piers in the larger resorts thus began to cater for a widening variety of tastes in the 1880s and especially the 1890s. Brighton's West Pier, though it retained its status as the town's fashionable pier, became famous for its sideshows, which included in 1890 a flea circus (featuring such tableaux as 'Cannon fired by a flea' and 'A morning drive with carriage drawn by fleas with fleas inside'). More widely socially acceptable, perhaps, was the real cannon which was 'fired precisely at noon every day (when sunny) by the sun's rays directed on to the touch-hole by a burning-glass'. Opulent Eastbourne found room on its expanded pier for a rifle saloon and an American bowling saloon; and mutoscopes or 'what-the-butler-saw' machines were appearing on the piers in most if not all of the larger resorts during the 1890s. More reputably again, there were athletic feats of swimming and diving by 'professors of natation', and most piers made a modest income from the issue of permits to anglers. Roller-skating, as at Southend, was another possible option. The full programme of an Edwardian pier in a popular resort, taking on board all the developments of the past twenty years, could make impressive reading. Thus Morecambe's Central Pier, advertising in the resort's official guide in 1906, offered (for a basic twopence admission to the pier and sixpence to the pavilion), the following catalogue of delights: 'The finest pier pavilion in the world'; dancing all day; two bands daily; morning promenade concerts at 11; concert and variety entertainments daily, at 2.30 and 7.30 ('All the leading stars, no duds'); sacred concerts on Sundays, at 3 and 8 p.m.; 'Grand Aquatic Performances and Sensational Diving into the Sea'; and daily steamers to Blackpool, Southport, Douglas and 'other places of interest'.

The continuing importance of steamers was general. Not only did they bring visitors, in competition with the trains, to the resorts of the Thames estuary, East Anglia and the Bristol Channel, as well as to the Isle of Wight and the Isle of Man (though here Douglas harbour took most of the arrivals), and (very importantly) down the Clyde estuary; they also provided day excursions which enhanced the attractions for visitors who were already established in the resorts. Thus Clacton-on-Sea was linked with Walton-on-the-Naze, Felixstowe, Harwich, Ipswich, Southwold, Lowestoft, Gorleston and Great Yarmouth, by a regular service from the 'Belle' steamer

Scarborough Pier, completed in
1869: an early photograph taken
from the castle.

line; and there were also boats to and from Southend, Gravesend and London itself. Bournemouth pier offered a regular service to Swanage, Southsea, Weymouth and the Isle of Wight, with longer summer trips to Torquay, Dartmouth, the Channel Islands and even Cherbourg, timetabled to return in time for the evening entertainments. Blackpool's piers ran steamer trips to the Lake District, using their own vessels; and Belfast and North Wales were also frequent destinations. In this way the piers provided additional flexibility in holidaymaking, and added to the many novelties and unusual experiences promised by the seaside. All this activity was still in full swing right up to the outbreak of the First World War.

The persisting importance of the original purposes of the promenade pier must not be forgotten, however. A Southport guide-book of the turn of the century provides a suitable reminder. The view from the pierhead is given pride of place in the Pier Company's own advertisement: 'A fine Panoramic View may be had, extending to Black Coombe in Cumberland, Lytham, the Estuary of the Ribble, the Lancashire Hills, the fine Promenade of Southport. . . the Channel of the Mersey, the Cheshire Hills, the Ormeshead on the Welsh Coast, and to the front of the spectator is the Open Sea'. The text praises the sensation of 'being upon the deck of a ship at anchor in smooth water. . . You have the breeze and the sea, without the sickness and rolling or pitching of a vessel. . . It is there the girls get the bloom on their cheeks again, and the pale faces of the town a tinge of the sun'. This clichéd guide-book prose perpetuated the commonplaces of popular expectation about the value as well as the pleasure of the promenade pier, and they remained important alongside the more sophisticated entertainments which were appearing in the later nineteenth century.

Piers continued to be built in a wide range of resorts between the turn of the century and the First World War, and pavilions continued to be added. At one extreme, little resorts like Minehead were still acquiring unostentatious promenade piers; and at the other end of

the spectrum came the Palace Pier at Brighton, which replaced the old Chain Pier. It was opened in 1899, and proved an instant success. Clifford Musgrave describes it as 'a highly attractive creation, with its golden oriental domes and delicate filigree ironwork arches, outlined at night in electric lights, that were inspired by the Pavilion's fretted stone lattices'. Architectural commentators endorse the enthusiasm of this patriotic local historian and celebrant of the joys of Brighton: to Kenneth Lindley 'it is considered by many to be the finest pleasure pier ever built'. It soon acquired a theatre, followed by a bandstand and a winter garden. This was, perhaps, the apotheosis of the pleasure pier; but the Grand Pier at Weston-super-Mare, built in 1903–4, ran the Palace a good second, with its domes, pavilions, hall, bandstand, shops and kiosks. The piers in the larger holiday centres were almost becoming full-scale resorts in their own right, with enough to keep a day-tripper happy for the duration of the visit.

It was still not all plain sailing for pier promoters, however, and it may be significant that each of these spectacular and expensive piers was threatened at an early stage by financial difficulties. The Palace Pier at Brighton had taken over the Chain Pier in 1891, and when the latter was washed away in a storm in 1896 the company had to find £6000 compensation for damage to the West Pier from the Chain Pier's wreckage, as well as £1500 for damage to Volk's Electric Railway and £2000 to repair their own half-completed structure. This forced them into liquidation, and a new company had to be formed to carry the work on. Weston's Grand Pier had been nearly twenty years in the making: powers had first been obtained in 1884, but a second Act of Parliament was necessary in 1893 after the original powers lapsed, and Weston's local authority became so convinced of the promoters' inability to proceed that it stepped in and sought leave in 1896 to build the town's second pier itself. This was opposed by both the Grand Pier company and by the existing Birnbeck Pier, and permission was refused, but only after a hard-fought battle before a parliamentary committee.

The opening of Blackpool's North
Pier in 1863, now so altered and
enlarged as to be barely recognizable.

Pier promotion thus remained a risky business, and there were many schemes that never came to fruition, 'piers that never were'. Blackpool might have had two additional piers, making five in all, by the early twentieth century. Two widely differing schemes bit the dust in 1898. The Blackpool Palatine Promenade Pier would have been the town's largest in area, though only 650 feet long, with an immense complex of buildings on a broad deck. There were plans for winter gardens, refreshment rooms, photographic studios, sideshows and a concert room, and the whole was to be surmounted by an enormous dome, ninety feet high. It was scuppered by furious local opposition from rival entertainment interests and people whose sea view would have been blocked, and it was also presented as a traffic hazard and a threat to the other piers' steamers. The Sea Water Company's proposals were more modest. Their pier was to be very small, and for promenading only; and it was designed in conjunction with a planned new pumping-house and pipeline to bring ashore the sea-water which the company supplied to baths and hotels, and sold in barrels to customers inland. But it would have invaded the most socially select area of Blackpool's promenade, and the wrath of the Corporation and the residents descended on the Company. There were suggestions that, once erected, the pier would soon be expanded and provided with amusements calculated to attract excursionists; and again, the plans were refused. Contested parliamentary hearings were expensive, and the syndicates involved in these speculations must have lost heavily by their defeat.

Changes to existing piers were also increasingly likely to encounter opposition from local authorities, rival vested interests and local residents by the late nineteenth century. Here again, Blackpool provides interesting examples. The South Jetty, or 'People's Pier' as it had become known, was especially controversial. In 1891 attempts to introduce 'dancing on a purpose-built dancing platform' at the shore end of the pier, to augment the existing 'penny hop' at the seaward end, were opposed by the police, local residents and the local authority. The main objection was to the noise and obstruction from the heavy-footed dancers and their bands, which made up in energy what they lacked in subtlety, and often began at 6 a.m. to cater for the earliest day excursion arrivals. Some of the lodging-house keepers near the pier were trying to keep their established middle-class visitors, and argued that the new proposals would very soon prove to be the last straw. There were also complaints about the pier's new buildings, which were held to obstruct the views of the Welsh mountains, which sea-front landladies had tried to make into a selling point. The pier company lost this argument, although they were allowed to introduce dancing 160 yards from the shore, at a safer distance from their antagonists. Seventeen years later, the South Jetty was in trouble again for introducing a Grotto Railway, which was thought to be the prelude to the development of a 'thundering big pavilion' for the still-controversial dancing. In this context, it is not surprising that both the abortive pier proposals of 1898 were at great pains to stress that under no circumstances would dancing be allowed on their premises.

Blackpool's piers were at the forefront of innovation in catering for the rapidly expanding working-class market of late-Victorian and Edwardian times, when the whole economy of the resort was in transition, and fears and hopes alike ran high. It is not surprising that proposals for development were sometimes disputed, challenged and blocked. Similar events occurred in Brighton, and in other popular resorts. In part, these initiatives were responses to the growing competitive pressures on piers. The *Blackpool Gazette* in 1908 assumed that fear of competition lay behind the South Jetty's Grotto Railway; and if these pressures were being felt in the centre of the world's most successful working-class resort, next door to the main excursion railway station, they must have been much more worrying elsewhere, even though no other pier had quite so many rival 'pleasure palaces' to reckon with on its own doorstep. Even in the boom years around the turn of the century, a pleasure pier certainly did not amount to

The West Pier, Brighton, soon after
its opening in 1866. Most of what we
see here still exists, with the addition
of a large pavilion in the centre.

anything resembling a licence to print money. Two aspects of pier finances stand out in this period. The major projects were being financed increasingly by consortia from London and the big industrial towns: local money was much less important than it had been in the first great wave of pier-building. Secondly, despite the interest being shown by outside capitalists, some pier companies were falling into serious financial difficulties and having to be rescued by local government. Southend was the pioneer in this respect, and the local authority's reconstruction of the pier in 1888–90, after several years of vacillation and dispute, more than trebled the annual receipts in three years. By 1909 the pier had brought in profits of £40,000 to set against the rates. This success story was widely diffused in other resorts: it was used at Great Yarmouth, for example, as an argument in favour of doing something similar to counter the dilapidation of the Wellington and Britannia Piers, both of which dated from the 1850s. Great Yarmouth, indeed, became one of several municipalities to become pleasure pier owners and operators between 1890 and 1914, joining Bournemouth, whose worm-eaten pier had been replaced by the local authority in 1880, as well as Southend. Other pier acquisitions took place at Clevedon, Ilfracombe, Bognor, Boscombe, Rhyl and Torquay. This unexpected extension of municipal responsibility reflects the perceived importance of piers to resort economies: it was dangerous to try to manage without one, or to make do with a sub-standard example. It also reflects the enterprise of seaside local government more generally, as local authorities stepped in where private enterprise feared to invest or to persist, providing amenities which were vital to the attractiveness and competitiveness of their towns.

By 1914 the golden age of pier-building was practically at an end. The pleasure pier itself had proved to be an astonishingly versatile and adaptable vehicle for popular enjoyment. From providing a deliberately exclusive promenade to separate the socially aspiring promenader from the great unwashed, and to promote health, gentle exercise and the cult of the picturesque, the pier had come to cater for the whole range of late-Victorian and Edwardian holiday tastes. The pressures for change were strongest in the large resorts close to the main population centres, of course; and the original ethos of the promenade pier survived almost unimpaired in some of the smaller places. Where there was more than one pier in a resort, too, the market could be divided to some extent, with one providing more raffish attractions than the other, as in Blackpool (where the new Victoria Pier of 1891 at South Shore was even more genteel than the mid-Victorian North Pier) and Brighton. But the keynote, above all, was adaptability to the changing demands and opportunities of an expanding and increasingly complex holiday market.

During the inter-war years, with their complex cross-currents of prosperity and unemployment (exemplified in the fact that Blackpool prospered while Lancashire's industries languished), the accent was on further embellishment and diversification. Pier construction dried up, but investment remained buoyant, though profitability was still sometimes problematic. The New Palace Pier, St Leonard's-on-Sea, illustrates these trends well. Originally constructed in 1889, it was effectively rebuilt in 1933. When it was put up for auction in 1938 it had an enormous deck area of 62,500 square feet – more than four times that of the largest Blackpool pier of the late 1890s – and it offered a remarkable range of attractions. It had a ballroom which could accommodate a thousand dancers on its sprung maple floor, a café, and in keeping with the trends of the times, a sun-lounge. Its amusement arcades were (said the auctioneers) 'reputed to comprise the most extensive and modern range of Automatic Machines on the South Coast', and bars, shops, kiosks and stalls proliferated. It had dodgems, skittles, darts and shooting ranges. Like the Victorian and Edwardian piers, too, it was liberally festooned with advertising posters and displays. But in terms of what it provided, it went far beyond the most adventurous of its pre-war precursors. Its management was rewarded with a million visitors in 1936, al-

ST. LEONARDS PIER.

The New Palace Pier, St. Leonard's-
on-Sea: opened in 1891 with an
ornate pavilion capable of holding
seven hundred people, but
demolished in 1951.

though its finances were extremely muddy and its profitability was, at best, very doubtful.

Similar trends were evident elsewhere, although we still know too little about seaside resorts between the wars to generalise with real confidence, and this applies especially to the important question of whether piers made profits, where, when and to what extent. Clacton's pier acquired (mostly between the wars) a dance hall, a switchback, dodgems, a children's railway, a swimming pool, a helter skelter and even a zoo. Eastbourne's pier added a concert pavilion in 1924 to the theatre which had been included in the reconstruction of 1901. Mostly, the additions and innovations involved going further down-market to attract the custom of growing numbers of day-trippers, and of a middle-class visiting public which was fast losing the last vestiges of Victorian staidness. Thus open-air dancing by the pier entrance was introduced at Brighton's Palace Pier and even at Bournemouth, while in 1927 the Palace Pier acquired a new and particularly titillating selection of 'what-the-butler-saw' machines, with such titles as 'Bedroom Secrets', 'Artist and Model' and 'Parisian Can-Can'. The range of entertainment permissible on Sunday was also extended in many resorts, as the tyranny of the obligatory 'sacred concert' became less universal. The steamer traffic, on the other hand, was sharply reduced, especially the day excursions. Many steamers were called up for service in the First World War and never returned, while sandbanks and changing tidal patterns also made life difficult in some resorts. At Blackpool, a combination of these factors led to the effective disappearance of steamer trips from the piers' menu of attractions in the 1920s.

Piers were unfashionable among the arbiters of taste in the inter-war years, as a reaction against all things Victorian gathered momentum. Harold Clunn, claiming in 1929 that Brighton's West Pier 'is probably the most elegant in Great Britain', felt obliged to make the immediate disclaimer that 'opinions on the merits of such structures as pleasure piers and railway stations vary enormously, and . . . this remark is merely intended by the writer as a personal expression of opinion'. Malcolm Muggeridge remarked that a pier was usually 'a kind of pasteboard Taj Mahal'; and the characteristic mock-Orientalism of Brighton Pavilion and piers alike came in for disapproval and mockery. But their popularity among holidaymakers at large continued to burgeon, despite the rise of new modes of holidaymaking and new holiday destinations during the 1920s and especially the 1930s.

The Second World War saw most of the pleasure piers, especially on the east and south coasts, deliberately breached in order to obstruct invasion: though why Hitler's army should have wanted to invade Britain along Clacton Pier is a mystery to those of us who are not military strategists. They were restored in 1946, to face an uncertain future, as competition of all kinds intensified and the cost of maintaining increasingly elderly structures began to escalate. There remained, and remain, outposts of popularity and optimistic investment, especially at Blackpool; but the decline and destruction of the seaside pier has gathered momentum over the last quarter of a century.

The vulnerability of piers to storm, collision and fire is nothing new. Piers on the east coast, especially, were regularly breached by storm-driven vessels or damaged by wind and tide in the 1870s and 1880s, and the threat from the elements has continued unabated. Even in the 1890s, damage was not automatically repaired in the smaller and less affluent resorts. Withernsea lost its pier to storms in 1880 and 1893–4, and there was insufficient wherewithal and enthusiasm to replace it. Coatham pier, near Redcar, was similarly not made good after a disastrous collision in 1898; but here two adjoining resorts had acquired rival piers which were more than a limited visiting public could support. Repairs and reinstatement were usual until recent years; but the casualty rate has grown as the irrelevance of piers to current holiday preoccupations has become increasingly apparent. Impressive levels of investment have continued in recent years at Blackpool, while Southport's pier has been protected, paradoxically, by

The sad spectacle of Coatham Pier, near Redcar, after it had been breached by a collision in 1898, and just before it was demolished the following year. It had been built in 1873.

the retreat of the sea which was its original reason for existing: it now strides purposefully across a municipal park, stretching a tentative toe into the water only towards high tide and at the very end of its enormous length. But elsewhere, the portents are gloomy, despite recent re-evaluations of the pier as architecture, feat of engineering design and epitome of the social history of an age which increasingly evokes nostalgia. In too many cases, the cost of repair and restoration have been deterring to companies which no longer have local roots and interests. The Victorian sewers beneath our cities, which have begun to show their age in similarly expensive ways, will have to be replaced: the piers are more attractive, but also more expendable. It is hard to escape the feeling that many of the photographs which follow are in some sense valedictory.

As we say farewell to so many of our piers, we should remember how important they have been to past modes of holidaymaking: and in the recent past, at that. As James Laver remarked in 1934, piers became 'in the end no longer mere elongated platforms, but, expanding at the tip and crowning themselves with pavilions, grew cafés and theatres, and dance halls, a microcosm of the larger world of amusement on shore'. Just so; and Laver went on to point to the less formal manifestations of the seaside pier, pointing to his own childhood fascination with 'the substructure, that forest of iron pillars, like trees with interlocking branches, a mechanistic, cubic forest, the more mysterious for the fact that it was often submerged'. Friends who attended dances on Morecambe's Central Pier in the late 1960s have mentioned other attributes of this nether world, as an ideal (if not very comfortable) secluded spot for exploring the delights of the opposite sex. Seaside piers were so successful, for so long, because they were so many things to so many people; and Laver draws attention to the way in which they combined elements of the cultured and controlled with a raw proximity to the beauties and savagery of the untrammelled natural world. Artifice and vulnerability, passion and security: the seductive, contradictory combinations could be multiplied. But

they have not been sufficient to preserve the seaside pier from the hard actualities of the changing leisure world and economic stringencies of the later twentieth century. The decline of the seaside pier is a tempting metaphor for the steady erosion of brash, ironclad Victorian certainties: it symbolises the demise of a cohesive culture and its replacement by storm-ridden diversity. The pier was able to adapt to these processes, up to a point; but in most places, for most purposes, it has now been left behind. Its past has vitality and social resonance: its future, sadly, may belong in the world of the museum, the preservation society and the professional purveyor of nostalgia.

John K. Walton.

Further Reading

Here is a list of readily available books which will help to provide further background information and ideas about the history of the seaside in general and piers in particular.

D. Cannadine, *Lords and Landlords: the Aristocracy and the Towns, 1774–1967* (1980): Eastbourne.

D. Cannadine (ed.), *Patricians, Power and Politics in Nineteenth-Century Towns* (Leicester, 1982): chapters on Southport and Bournemouth.

Yvonne Cloud (ed.), *Beside the Seaside* (1934): includes Malcolm Muggeridge on Bournemouth and James Laver on Blackpool.

Harold Clunn, *Famous South Coast Health and Pleasure Resorts* (1929).

E. W. Gilbert, *Brighton, Old Ocean's Bauble* (1954, reissued 1976).

Kenneth Lindley, *Seaside Architecture* (1973).

J. Lowerson and J. Myerscough, *Time to Spare in Victorian England* (Hassocks, 1977): chapters on the Sussex coast.

Clifford Musgrave, *Life in Brighton* (1970): full of loving and entertaining detail.

J. A. R. Pimlott, *The Englishman's Holiday: a Social History* (1947, reprinted 1976): a wonderfully entertaining book, and years ahead of its time in research and content.

E. Sigsworth (ed.), *Ports and Resorts in the Regions* (Hull, 1980): several chapters on individual resorts.

J. K. Walton, *The Blackpool Landlady: a Social History* (Manchester, 1978).

J. K. Walton, *The English Seaside Resort: a Social History, 1750–1914* (Leicester, 1983).

J. K. Walton and James Walvin (eds.), *Leisure in Britain, 1780–1939* (Manchester, 1983; paperback edn., 1986): chapters on Blackpool, Bournemouth and Ilfracombe.

James Walvin, *Beside the Seaside* (1978).

D. S. Young, *The Story of Bournemouth* (1957).

BRIGHTON. PALACE PIER.
Decorative ironwork and deckchairs
facing the sun. Through the slats of the
floor the visitor watches the sea rising
and falling – epitome of the seaside
experience.

BRIGHTON. PALACE PIER.
Still one of the most impressive of
British piers, this was built after
the old Chain Pier collapsed in a
storm in 1896. It is 500 metres long
and is carried on 368 supports.

BRIGHTON. PALACE PIER.
In the evening light the pier looks
like same strange fantasy from
Victorian science fiction.

Through the giant legs of the Palace
Pier one sees the ghostly outline of
the West Pier.

BRIGHTON. PALACE PIER.
Summer offers the visitor a selection of
cosmetics and perfumes, famous names
from haute couture mingling unexpec-
tedly with seaside banalities.

Oriental symbolism was a feature
of many piers – onion domes and
horseshoe arches hung with coloured
lights.

BRIGHTON. WEST PIER.
Designed by the great engineer
Eugenius Birch and opened in 1865,
the West Pier is the Cinderella of
British piers. It closed in 1975
and since then its condition has
deteriorated. Millions of pounds are
needed for its restoration but the
money seems likely to be raised.

A sad memorial to the power of the wind and water. The pier is now closed as a dangerous structure.

BRIGHTON. WEST PIER.
Victorian ironwork combined frivolity
and functionalism in a way that seems to
elude modern designers. It takes little
imagination to see troupes of ladies with
parasols and hats parading under these
canopies.

Ruin of a bygone age. The West Pier
now seems like a forlorn outpost
of elegance in a world of faceless
concrete.

BRIGHTON. WEST PIER.
Decaying splendours, but still capable of
revival. The concert hall
(top right) was added in 1916.

Bingo was the most popular game in
Britain during the 1960s. The West Pier
did its best to keep up with the fashions
until its closure in 1975.

HASTINGS.
Another structure by Eugenius Birch,
the most prolific of pier designers, but
so rebuilt that little of his work remains.
It has been compared to a giant insect
creeping with its thousand legs across
the sands.

HASTINGS.
From doughnuts to T-shirts
– everything the holiday-maker thinks
he needs. The building in the centre is
the Triodome, built as late as 1966.

The end of the pier is now occupied by a new pavilion that replaces the original destroyed by fire in 1917.

EASTBOURNE.
Opened in 1872, designed by Eugenius
Birch and not drastically altered since,
though much of it had to be rebuilt after
a storm in 1877. The way the cast-iron
buildings (seen in detail below left)
overhang the sides is particularly
charming.

As with most piers, large new structures have been built here during the hundred years of its existence. The theatre at the very end dates from 1901.

EASTBOURNE.
Looking out to sea from the end of the
pier, the point to which every visitor is
drawn. Eastbourne is one of the sunniest
places in the British Isles.

Foreshortened view (taken from the
seaward end) of the music pavilion
built in 1925. It is now an amusement
arcade.

SOUTHEND-ON-SEA.
Opened in 1890 and two kilometres in
length, Southend Pier served as a landing-stage for steamers. It was severely
damaged in 1986 when a coaster col-
lided with it.

Ghostly lights in the semi-darkness
signal the passage of a train, which
formerly ran the entire length of the
pier.

CLACTON-ON-SEA.
The pier was opened in 1860 and
lengthened in the 1890s. None of the
fairground structures is very old.
Clacton has a whole galaxy of these.

CLACTON-ON-SEA.
Clacton boasts such marvels as the
largest suspended swimming pool in
the world, a miniature race-track and
the seductive challenge of Mr Muscle.

A four-lane slide, from the heights
of a Spanish castle to the deck just above
the water.

WALTON-ON-THE-NAZE.
A pale shadow of the great Victorian
piers, Walton seems a cold, functional
structure of the modern age.

Long planks lead the eye into the distance under a lowering sky. Walton once had a railway.

GREAT YARMOUTH. BRITANNIA PIER.
The present nondescript pavilion and
theatre were built as recently as 1958. It
is interesting chiefly as
a barometer of popular taste in
entertainment.

England's answer to Las Vegas.
Once it was performing fleas; now it
is Adrian Mole, Jim Davidson and
Showaddywaddy, 'the greatest
Rock'n'Roll show ever.'

GREAT YARMOUTH. WELLINGTON PIER.
The substructure dates from 1854, but
the theatre is a flamboyant
exercise in the Art Deco style.

GREAT YARMOUTH. WELLINGTON PIER.
Clever ligthing gives an air of magic to
the pier by night that it lacks in the
daytime.

Season of Stars. The pier theatres
are still a regular circuit for the
hopefuls of the entertainment
industry. Some have already broken
into the great world of television.

CROMER.
The ramp, just visible at the end of the
pier, is a slipway for lifeboats. The blue
plastic in the foreground is an inflatable
castle for children to jump on. On the
pier itself you can buy any size of panda,
or shelter from the wind with a nice cup
of tea.

CLEETHORPES.
Built in 1875, Cleethorpes Pier was
originally more than twice as long. The
pavilion, dating from 1903, was then in
the middle. The seaward end was demolished during the Second World War
for military reasons.

CLEETHORPES.
'Pier 39' is billed as a 'supermodern'
disco, part of an English chain,
bringing the old structure into the age of
Hi-tech.

72

Morning. The disco is asleep. Peace
and quiet return to the beach. For
the local inhabitants piers have
always been a doubtful blessing.

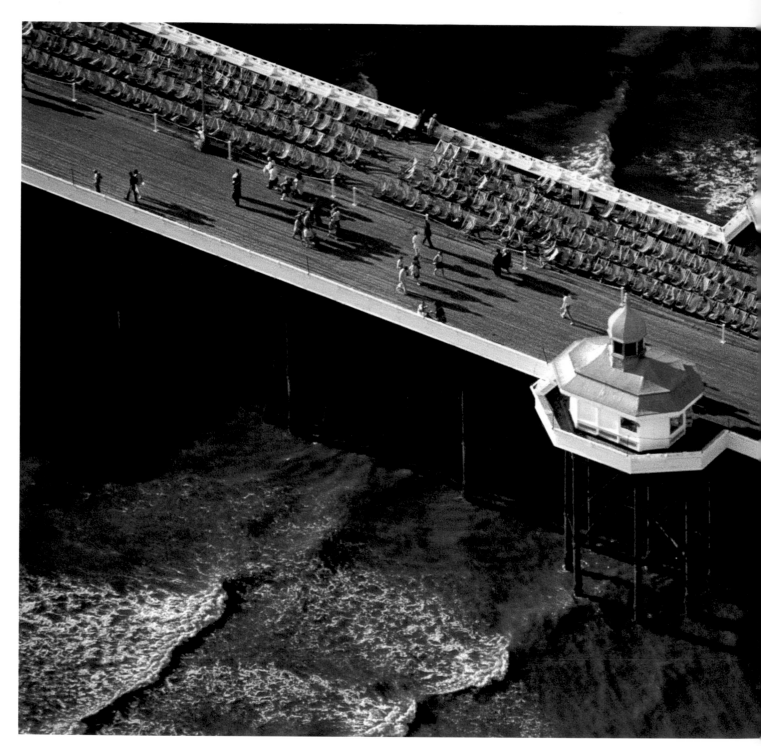

BLACKPOOL. NORTH PIER.
This is one of Eugenius Birch's
creations, opened in 1863 and
enlarged ten years later. Its wide deck
holds 3,000 deckchairs, here lined up
to face the sun.

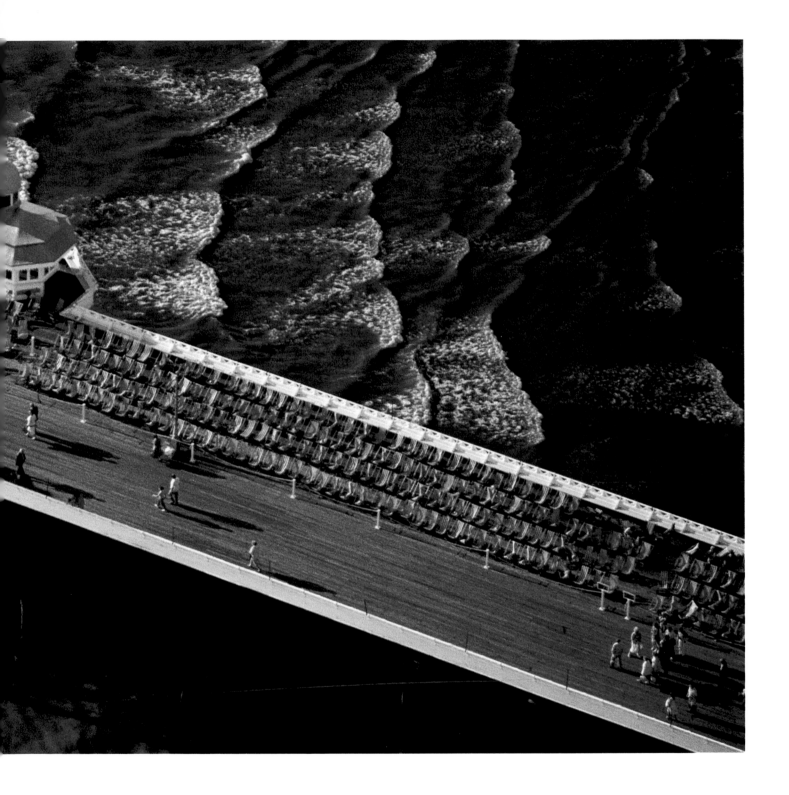

BLACKPOOL. NORTH PIER.
Towards the seaward end the pier swells
into vast platforms holding amusement
arcades and a large theatre.

Some of Birch's Victorian iron pillars
which still support the structure,
made sinister and mysterious by a
hundred years of rust.

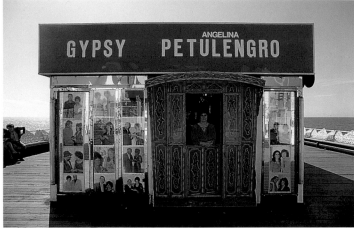

BLACKPOOL. NORTH PIER.
Splendid iron seats for which the pier is
famous; Gypsy Petulengro, fortune-
teller; the two other Blackpool Piers,
Central and South, seen through the
struts of the North Pier; and another
airview.

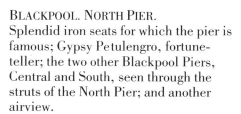

Looking north, one gets a good view
of Central Pier, opened in 1868.
Central Pier has the greater
collection of entertainments — from
helter-skelters to restaurants, from
dance-halls to boutiques.

ST ANNE'S-ON-SEA.
Opened in 1885, St. Anne's is still one
of the most exotic of British piers, with
its Chinese-style kiosks and elegant
ironwork.

ST ANNE'S-ON-SEA.
Part of the landward end has been screened
to give protection from the rigours of the
Irish Sea. Like many Lancashire resorts,
St. Anne's has been affected by the recession
of the coast, leaving wastes of bare sand
where there was once open sea.

LLANDUDNO.
For the connoisseurs this is probably
the best of all piers and certainly the
best preserved. It was built in 1876 and
steamers still operate from the end of it.
The kiosks and pavilions, though mod-
ernized, retain an authentically
Victorian atmosphere.

MUMBLES.
Very little of the original pier of 1898,
survives. Today it is virtually
a fairground built on water.

A touch of Disneyland comes through
in the red dinosaur and the gorilla
whose arms serve as a see-saw.

PENARTH.
The old pier was destroyed in 1931 and
has been twice rebuilt, including a
pavilion in a festive Art Deco style. The
Waverley Company operates a cruise in
the Bristol Channel which uses an old
paddle-steamer.

One of the less arduous pleasures of
pier-life is the angling competition,
a sport that attracts large numbers.
The flag floating over the roof is that
of Wales.

CLEVEDON.
With its long legs and wide spans, Clevedon, opened in 1868, is one of the masterpieces of Victorian engineering. One whole section collapsed during a load-bearing test in 1970 and since then the end, with its little pavilion, has stood like an abandoned island. The case has aroused more interest than any other pier, and after a fight by conservationists the money has been found to repair it.

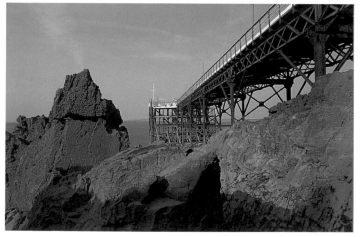

WESTON-SUPER-MARE. BIRNBECK PIER.
This pier, another work by Eugenius
Birch, connects the shore with
Birnbeck Island and from there goes
on into the open sea. It opened in
1867 and was for many years one of
the most popular piers in the
country, its attractions including
a water-chute and a switchback.

WESTON-SUPER-MARE. GRAND PIER.
In 1904 a second pier was built at
Weston, shorter than the Birnbeck Pier
but amply provided with amusements.
The ingenious lighting lends it a
glamour to which the architecture alone
cannot aspire.

The glass-roofed pavilion at the end of the pier was built in 1933. It is the biggest covered space on any pier, as large as a railway station. It contains a funfare and all sorts of other amusements.

96